W9-CED-840

Hawk Highway in the Sky

Watching Raptor Migration

CAROLINE ARNOLD

HAWK HIGHWAY IN THE SKY

WATCHING RAPTOR MIGRATION

Photographs by
ROBERT KRUIDENIER

A GULLIVER GREEN BOOK
HARCOURT BRACE & COMPANY
San Diego   New York   London

Gulliver Green is a registered trademark of Harcourt Brace & Company.

Library of Congress Cataloging-in-Publication Data
Arnold, Caroline.
Hawk highway in the sky: watching raptor migration/by Caroline Arnold;
photographs by Robert Kruidenier.
p.   cm.
"A Gulliver Green book."
Includes index.
Summary: Provides information about hawks, eagles,
and falcons and efforts to study them, especially the HawkWatch International
Raptor Migration Project in the Goshute Mountains in Nevada.
ISBN 0-15-200868-3   ISBN 0-15-200040-2 (pbk.)
1. Hawks—Juvenile literature. 2. Eagles—Juvenile literature.
3. Falcons—Juvenile literature. 4. Hawks—Migration—Juvenile literature.
5. Eagles—Migration—Juvenile literature. 6. Falcons—Migration—Juvenile literature.
7. Bird banding—Nevada—Goshute Mountains—Juvenile literature.
[1. Hawks. 2. Eagles. 3. Falcons. 4. Birds—Migration.]
I. Kruidenier, Robert, 1946– ill. II. Title.
QL696.F32A76   1997
598.9'16—dc20   95-51213

Designed by Linda Lockowitz and Lisa Peters
Text set in Simoncini Garamond
First edition
A C E F D B
A C E F D B (pbk.)

Printed in Singapore

Gulliver Green® books focus on various aspects of ecology
and the environment, and a portion of the proceeds from the sale of these books
is donated to protect, preserve, and restore native forests.

Photo on front jacket/cover: A red-tailed hawk
Photo on back jacket/cover: A golden eagle
Photo opposite title page: A male American kestrel

Acknowledgments

I HAVE WATCHED birds all my life, but I have never seen so many hawks and falcons as during my visit to the HawkWatch International Raptor Migration Project in the Goshute Mountains in Nevada. I want to thank all of the staff and volunteers for welcoming me and sharing their expertise and passion for these beautiful predatory birds. Thanks also to the U.S. Bureau of Land Management, Elko District Office, for its constant support of that project. I am particularly grateful to Steve Hoffman (pictured above), the founder and president of HawkWatch International, Inc., for his special assistance and for reading and criticizing the manuscript. I also thank Adele Binning, developer of the traveling exhibit "Hunters of the Sky" at the Science Museum of Minnesota, for introducing me to Robert Kruidenier, a devoted HawkWatch volunteer and superb photographer.

HawkWatch International is a nonprofit organization devoted to the study and protection of migratory raptors in the western United States. Further information about HawkWatch International projects in the Goshutes and at other locations and about their Adopt-a-Hawk program is available by writing to them at P.O. Box 660, Salt Lake City, Utah, 84110, or by calling 1-800-726-4295, or by visiting them at their World Wide Web page, http://www.vpp.com/HawkWatch. Although this book focuses on the observation and banding of migrating raptors in the Goshute Mountains, people monitor migrating raptors at many other sites along bird migration routes in North America. (Most sites are just for observation and do not include trapping stations like those in the Goshutes.) Look on page 47 for a map and list of a few locations for hawk watching in North America. You can find out about other places to watch raptors by contacting your local Audubon Society or by writing the Hawk Migration Association of North America, P.O. Box 3482, Lynchburg, Virginia 24503.

HIGH IN THE SKY, a hawk spreads its wide wings and circles over a remote peak in eastern Nevada. When it spies a fluttering pigeon in a clearing below, it dives earthward, anticipating a tasty meal. But just before reaching the ground, the hawk slams into a net and becomes trapped in a tangle of fine cords. The pigeon is quickly forgotten as the bird struggles to get free. Biologists capture the hawk, remove it from the net, and proceed with a series of measurements of its wings, talons, beak, and body. They weigh and examine the bird and place an aluminum band on its leg. The entire process takes about fifteen minutes, after which the scientists release the hawk with a quick boost into the air. The hawk takes off, seemingly undisturbed by this brief interruption in its day, and continues flying to its winter home far to the south.

Nearly four thousand migrating hawks, eagles, and falcons are banded each fall in the Goshute Mountains of eastern Nevada by scientists and volunteers working with an organization called HawkWatch International. HawkWatch observers count another ten to twenty thousand birds as they fly overhead. The study of birds in the Goshutes, along with information gathered at other hawk migration sites in North America, is helping us to learn more about these impressive birds of prey and their annual journeys between their winter and summer homes.

Powerful wings lift this bird into the air
as it is released after being banded.

HAWKS, EAGLES, AND FALCONS are predatory birds called raptors. Raptors have strong grasping feet, sharp talons, a hooked upper beak, and excellent vision—qualities that make them superb hunters. Owls are raptors, too, but they specialize in flying and hunting at night whereas hawks, eagles, and falcons are active by day.

Raptors, like other hunters, depend on an adequate supply of prey for their survival. Their prey, in turn, feed on smaller animals, insects, and plants. The interdependence of plants and animals in the environment is called the food chain, or the food web. Each part of the chain is linked to every other part. For example, plants need water, rodents eat plants, and many hawks eat rodents. Because raptors are at the top of the food chain, they reflect changes that occur elsewhere in the chain. In a drought year, for example, fewer plants grow, so rodents are less numerous. Then it is more difficult for hawks to find food and fewer survive. Monitoring the health of raptors is one way to gauge the general health of the environment. Yearly counts of migrating hawks, eagles, and falcons provide scientists with information they need to help protect and manage wild areas where the birds live.

ABOVE: *Examinations of captured birds help biologists to assess their overall health.*

RIGHT: *A young Cooper's hawk*

INCREASES AND DECREASES in raptor populations from year to year are a result of normal variations in weather patterns and other natural events. However, raptor counts conducted over many years help scientists to detect long-term trends that reveal more permanent alterations to the environment. As wild areas are destroyed by large-scale, intensive agriculture, overgrazing by livestock, clear-cut logging, and urban development, the animals that once lived there have nowhere to go. Other human-caused environmental hazards to raptors include illegal shooting, collisions with fences and utility lines, electrocution on power lines, poisoning from lead shot, and the harmful effects of pesticides.

In the 1950s, studies of hawk migration led to the discovery of the damaging effects of the pesticide DDT in the environment. When an animal eats food containing DDT, the chemical is stored in its body. When raptors eat these animals, the pesticide is passed on to them. The DDT stored in each prey animal can accumulate to dangerous levels in the body of a predatory bird. DDT causes the eggshells of these birds to weaken and break before the baby birds are ready to hatch. The populations of many raptors plummeted when DDT was in use because the birds were not able to reproduce. Since 1972 the use of DDT and related compounds has been banned in the United States. Although most birds are now able to raise their young successfully, pesticides continue to be a problem for some raptors. DDT is still used in some Central and South American countries, where many migrating raptors or their prey spend the winter. New pesticides now being used in the United States, Canada, and Latin America might also prove to have harmful effects on raptors. Careful study of raptors may help scientists find out.

Peregrine falcons nearly became extinct because of DDT but have made a comeback with the help of captive breeding programs. These birds have been successfully reestablished in both urban and wilderness areas.

THIRTY-ONE SPECIES of hawks, eagles, and falcons live in North America. They are found in almost every type of habitat, ranging from open plains and dense forests to rocky deserts and occasionally even our biggest cities. Many raptors are large birds, and you can often see them circling high in the sky or perched on a pole or tree. Few sights are as thrilling as that of a red-tailed hawk or golden eagle soaring with effortless grace as it scans the landscape below for prey.

When birds migrate in fall and spring, you can visit many of the lookout sites where large numbers of raptors pass by, and you can join observers as they watch for birds flying overhead. Every fall hundreds of visitors climb to the top of the Goshute ridge to look for migrating raptors and to observe people trapping and banding birds. The opportunity to see these fascinating birds close-up helps us to appreciate and understand the diverse ways in which they live.

RIGHT: A young visitor gets a close look at a Cooper's hawk.

BELOW: Visitors to the Goshutes learn about a red-tailed hawk just before its release.

ABOVE: *Red-tailed hawks range in color from light to dark.*

RIGHT: *Adult red-tailed hawks can be easily identified by their large size and bright, russet-colored tail feathers.*

SCIENTISTS GROUP hawks, eagles, and falcons into two large families, the Accipitridae and the Falconidae. Birds within each family have some similar characteristics, but each species is unique and is adapted to its own way of life. The Accipitridae family includes all of the hawks, eagles, and kites, as well as the ospreys and harriers. The Falconidae family includes the falcons and the caracaras. (Look on page 46 for a list of North American day-flying raptors.)

Scientists classify hawks according to their body shape as either buteos or accipiters. The buteos have large, broad wings and broad, rounded tails. These large birds typically soar in high, wide circles as they hunt for food over prairies, fields, and other open spaces. The buteos' exceptional soaring abilities help them to cover large territories with relatively little effort.

Of the twelve species of buteos that live in North America, the red-tailed hawk is the most common and can be seen throughout the continent. These hawks often perch on tall poles or at the tops of trees, where they are easy to spot. The red-tailed hawk is the buteo caught most frequently in the Goshutes. Other buteos that pass over the Goshutes include the Swainson's hawk and the broad-winged, the rough-legged, and the ferruginous hawks.

ACCIPITERS, which are sometimes called true hawks, have shorter, more rounded wings than buteos and longer tails, features that make them better suited to moving quickly through the dense woodland areas where they live. Accipiters use their long tails for steering, and their short, strong wings give them power to chase prey through the trees. They seldom soar except during migration. Instead, they usually fly by flapping and gliding.

Three species of accipiters live in North America—the northern goshawk, Cooper's hawk, and sharp-shinned hawk. Cooper's and sharp-shinned hawks are the most frequently caught birds at the trapping stations in the Goshutes and account for about 80 percent of the total number of birds trapped and banded each year at that site.

TOP: *An immature male sharp-shinned hawk rests quietly for a moment before taking off.*

RIGHT: *An adult Cooper's hawk*

LEFT: *An adult northern goshawk. Like many hawks and falcons, goshawks have a ridge over each eye, forming a kind of "eyebrow" that helps shade and protect the eye.*

EAGLES ARE AMONG the most impressive birds trapped in the Goshutes. They are the largest raptors in the world, with wingspans that often measure more than seven feet. They can be recognized by their large size, wide wings, and wedge-shaped tails. Adult bald eagles have white heads and tails, whereas adult golden eagles have dark heads and tails. Both of these North American eagles migrate along the Goshute ridge, although golden eagles are much more common.

Kites, which get their name because their style of flight is similar to that of a kite floating in air, are found mainly in Central America and the southeastern parts of the United States. They do not migrate over the Goshutes.

The osprey is mainly a fish eater and is usually found close to rivers and lakes. It is seen rarely in the Goshutes.

The harrier, also sometimes called a marsh hawk, glides on long wings close to the ground as it searches for prey in grassy and marshy areas. Harriers frequently fly over the Goshute ridge.

ABOVE: Raptors range in size from the tiny sharp-shinned hawk to the huge golden eagle.

RIGHT: Harriers can be recognized by their slim bodies, long tails, and narrow wings.

ABOVE: *The merlin is a small, powerful falcon with a stocky, angular body.*

LEFT: *Kestrels are the smallest falcons and the most commonly caught falcons in the Goshutes.*

RIGHT: *A prairie falcon is released after being banded.*

FALCONS BELONG TO the Falconidae family and can be recognized by their narrow, pointed wings and long tails. They are extremely fast fliers and are able to catch and kill other birds in the air. Peregrine falcons, which feed almost exclusively on smaller birds, are able to attain speeds of up to two hundred miles per hour as they dive toward their prey.

Falcons that live in North America include the gyrfalcon, peregrine falcon, prairie falcon, aplomado falcon, merlin, and American kestrel, and all except gyrfalcons and aplomado falcons migrate over the Goshutes. Gyrfalcons, which are the largest falcons, live in Alaska and northern Canada. In winter they migrate to southern Canada and occasionally to the northern United States. The aplomado falcon, which formerly bred in the southwestern United States, is now found only in southern Texas, Mexico, and Central and South America. The crested caracara, another member of the Falconidae family, also lives in Central and South America as well as in parts of Florida and the southwestern United States. This large, long-legged bird inhabits open areas and is both a hunter of live prey and a scavenger of dead animals.

RAPTORS, LIKE MANY OTHER BIRDS, migrate, or travel, from one place to another to find food, places to nest, or a more comfortable climate. Migration is an instinctive behavior, and young birds know where to go even though they have never been there before. Migrating birds that breed in the Northern Hemisphere generally travel south in the fall and return north to their breeding grounds in the spring. (In the western United States, some of the raptors that breed in the Rocky Mountains and Great Basin desert migrate eastward to the Great Plains in winter.) Not all birds migrate. In some cases only young birds migrate. In other cases the birds migrate only in years when food is scarce. Migration behavior varies from one species of bird to another, and even within the same species.

Scientists do not completely understand why birds migrate or exactly how they find their way on journeys that are sometimes thousands of miles long. Some North American birds fly to the tip of South America and back each year, while others migrate just a few hundred miles. Scientists are still learning how birds find their way on these journeys.

One way that migrating birds find their direction when they travel is by looking at the angle of the sun in relation to the horizon and by using the sky as a map. Birds also have a kind of internal compass that detects magnetic forces within the earth. Some birds may use extremely low sounds, called infrasounds, made by thunder or wind to help them navigate. Other helpful guides for migrating birds are the large geographical landmarks that they see below them as they fly. Major landmarks for migrating raptors in North America include the coastlines of the Atlantic and Pacific Oceans and the Great Lakes, the Appalachian Mountains in the East, and the Rocky Mountains in the West. The Goshutes are one of a series of smaller mountain ranges located in the inter-mountain region between the Rockies and the Sierra Nevada.

A golden eagle spreads its wings over the Goshutes.

MIGRATING RAPTORS usually choose routes that enable them to fly with the least amount of energy. One of the best places to fly is along mountain ridges, where rising air helps keep the birds aloft. When moving air meets a mountain range, it pushes up and forms an updraft. Columns of warm upward moving air called thermals also form along mountain ridges and in the valleys between mountains. They can be up to two miles high. The wide wings of predatory birds are well suited to catching these rising air currents, and the birds ride them almost as if they were in invisible elevators. Once a bird has reached the top of an updraft or thermal, it then takes off on a long glide to the next one.

During migration season you may see a large number of birds rising in thermals together. A group of hawks or falcons flying together in this way is called a kettle. Broad-winged hawks and Swainson's hawks are species that often migrate in groups, and they sometimes form kettles of a hundred birds or more.

RIGHT: *Migrating birds, like this peregrine falcon, follow the Goshute ridge to avoid crossing the hot, dry desert to the east.*

BELOW: *Updrafts created along the Goshute ridge, which rises to nine thousand feet above sea level, help migrating birds conserve energy as they fly.*

As raptors migrate, they follow routes where food is available and where they can find safe perches for the night. An average daily distance for a migrating hawk is about 100 to 150 miles, although some birds fly 300 or more miles in a day. Birds generally fly directly south or north, but when they come to barren deserts or large bodies of water, they often change direction to go around them or to take the shortest route across. Flying across such expanses is dangerous because of the lack of food and resting places. The best hawk-watching sites are often located near natural obstacles or at places where several migration routes meet. At Cape May, New Jersey, in the United States, birds traveling south along the Atlantic Coast congregate over the last point of land before crossing the open water at the mouth of the Delaware Bay. At Hawk Ridge, near Duluth, Minnesota, raptors come together as they fly around the western tip of Lake Superior. In the intermountain region of the western United States, birds coming from a wide area to the north and northwest are funneled along the Goshute ridge, the first large mountain range west of the Great Salt Lake Desert. One of the best places in the world to see hawks and falcons is in Veracruz, Mexico, where migration paths from all across North America meet on the narrow strip of land between the Gulf of Mexico and the high Sierra Madre.

THE TIMING OF RAPTOR MIGRATION varies with the species and where the birds live. Fall migrating birds may not leave their summer breeding territories until the weather turns cold or food becomes scarce. In other cases, birds begin their journeys when internal body "clocks," triggered by shortening day lengths, tell them it is time to leave.

In the Goshutes, the first migrating birds appear in mid-August, and the last birds fly over by early November. Every day during this period, people scan the skies for migrating raptors. From sunup to sundown, trained observers sit on top of the Goshute ridge and count the birds flying overhead. Some birds fly so close that they can be seen easily with the naked eye. Others are mere specks in the sky and can be identified only with the help of powerful binoculars. The silhouette of the bird in flight helps an observer quickly discover its general type. Each species also has identifying marks and a characteristic flight style. With sharp eyes and years of experience, observers can almost always identify the species of a bird flying overhead, and they can often determine its age and sex as well.

In addition to hawks, eagles, and falcons, observers see migrating turkey vultures. These large birds are distinguished by the V-shape of their wings in flight. Although vultures are often seen with raptors and counted at raptor observation sites, they are in their own scientific family, the Cathartidae. Vultures do not have the sharp claws and strong beaks of raptors. They do not hunt live prey but feed only on animals that are already dead.

On an average day about two hundred raptors pass over the Goshute ridge, but at the height of the migration season, observers may count more than a thousand birds in a single day. The Goshute Mountain observation site is the busiest location for migrating raptors in western North America.

TOP: *Observers on top of the Goshute ridge cross-check their sightings with each other to confirm identifications.*

BOTTOM LEFT: *A written record is made of every bird that is sighted.*

BOTTOM RIGHT: *Working into the night, observers tabulate the results of their observations for that day.*

HAWK MIGRATION counts from sites across North America provide scientists with a yearly record of how many and what kinds of birds are migrating. To obtain detailed information about the birds, however, scientists must examine them close-up. This is accomplished by catching some of the birds. The HawkWatch site in the Goshutes is one of a number of places in North America where migrating raptors are trapped and banded.

The capture stations in the Goshutes are located near the top of the ridge in meadows or on open knolls that can be seen easily by birds flying overhead. Not all birds fly the exact same route, and the birds' flight patterns vary with the weather and wind direction. A variety of trapping locations makes it more likely that a bird will be attracted to one of them.

Raptors have extremely sharp eyesight estimated to be three to five times better than that of humans. From high in the air they can spot the movement of a small bird or rodent on the ground. As migrating raptors pass over the trapping stations, they are attracted to the fluttering of a pigeon that is placed near the traps. Scientists use non-native birds such as pigeons as lures to get the attention of raptors flying overhead. The hungry raptor dives toward the lure bird but instead of catching a meal, it finds itself enmeshed in the cords of a net. About 15 percent of the passing birds come down to the traps and get caught.

TOP: *A northern goshawk approaches a bow net,*

MIDDLE: *lands,*

BOTTOM: *and is caught.*

RIGHT: *A female kestrel is caught when her feet and wings become entangled in a mist net.*

Several types of traps are used to catch raptors. In the Goshutes, many of the smaller birds, such as sharp-shinned hawks and kestrels, are caught in the fine mist nets that are strung across the trapping area. Larger birds such as goshawks and red-tailed hawks are usually caught in circular bow nets on the ground or in small collapsing nets.

LEFT: *A trapper inside a blind pulls a string to close a bow net and catch a bird.*

BELOW: *A separate blind enables visitors to get a close-up view of the trapping process.*

TOP RIGHT: *An incoming bird does not usually see the fine cords of a mist net until it is too late to avoid crashing into it. Trappers need nimble fingers to disentangle a bird quickly.*

BOTTOM RIGHT: *A harrier is removed from a bow net.*

EACH OF THE TRAPPING STATIONS in the Goshutes also has a small shed, or "blind," that hides people from view while they wait for incoming birds. Raptors are naturally wary of people and avoid the trapping stations if they detect people nearby. Blinds are usually built near trees or bushes to make them as inconspicuous as possible. A covering of leaves and camouflage coloring also helps to make them look like part of the landscape. Trappers sit inside the darkened enclosure and peer out of a narrow slit as they scan the sky for approaching birds. They also watch the traps, and the instant a bird is ensnared, someone runs out to get it free. Experienced trappers know how to do this quickly without being injured by the raptor's sharp beak and talons. Trappers learn how to handle birds safely and do not need to wear gloves.

Trappers bring the captured bird into the blind and place it in a holding can. Airholes in the can allow the bird to breathe easily, and the confinement helps the bird feel calm and secure. The can also makes it easier for people to handle the bird safely and process it quickly. They record the bird's species, sex, and age along with the date, place, and type of trap in which it was caught. Then they assign a band number to the bird. This number, which is indented on a strip of lightweight metal, will identify the bird for the rest of its life. The circular band is clamped around the bird's right leg so that it can slide easily, but not so loose that it will catch on anything.

On days when large numbers of birds are trapped, the birds are released immediately after their bands are in place. Most of the time, however, the processing of each bird continues with a detailed examination and a series of up to ten different measurements.

ABOVE: A male kestrel is slipped into a holding can.

TOP RIGHT: A metal band is clamped onto a bird's leg while it is in the can.

BOTTOM RIGHT: Precise instruments provide exact measurements of each bird.

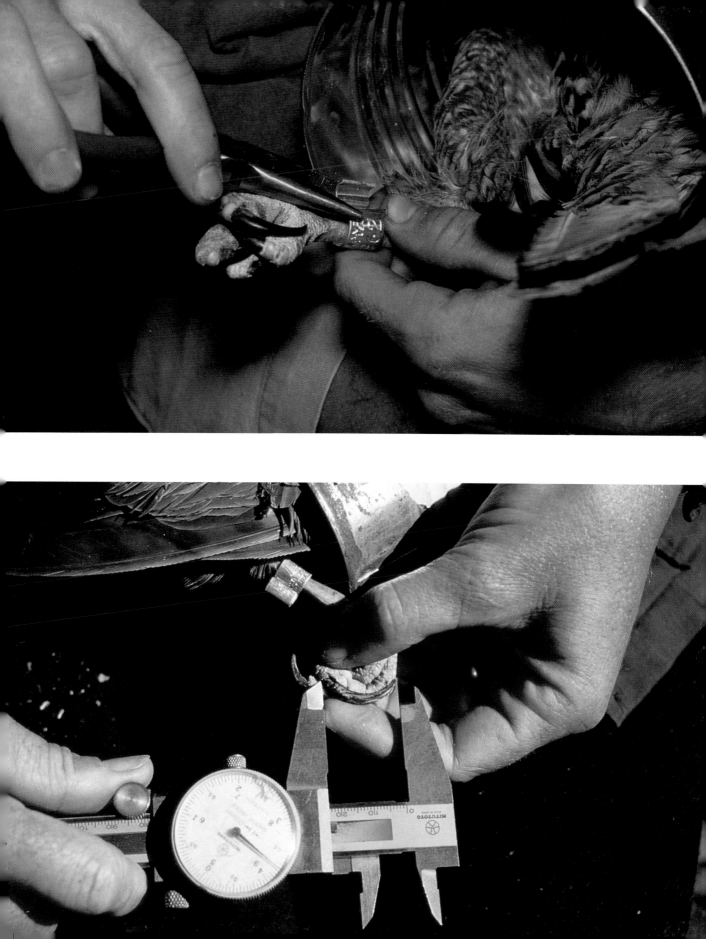

ONE OF THE FIRST MEASUREMENTS taken from each captured bird is its weight. Body weight is one indication of a bird's health, and it also helps to verify its sex. In some raptor species, males and females have different markings, but in others it can be difficult to tell the sexes apart except by their size. Raptors are unlike most other birds in that the female is larger than the male, typically by as much as one-third.

LEFT: *Birds are examined at the back of the blind and as each piece of information is obtained, it is recorded on a chart.*

RIGHT: *The adult male sharp-shinned hawk on the left is noticeably smaller than the adult female on the right. Although both birds have reached their full size, the female's red eye indicates that she is older.*

The advantages of size difference in raptors is a fascinating question that scientists debate and study. In general, large raptors hunt larger prey and small raptors hunt smaller animals. If a male and female of the same species hunt different size prey, they are less likely to compete with each other for food. Even though raptors are expert hunters, it is not always easy to locate prey, and birds often have difficulty finding enough food. Raptors are extremely protective of their food and reluctant to give it up, even to their mates. A larger female is able to dominate her mate, which makes it easier for her to take food from him when he is doing most of the hunting and she is sitting on eggs in the nest. She also can more easily cover and incubate the eggs. An advantage of the male's smaller size is that he uses less energy and thus needs less food. He is also more agile, which helps him to catch fast-moving prey.

RAPTORS HAVE SHORT, STRONG LEGS with powerful feet that are well adapted for catching and killing prey. The feet of most raptors have three toes facing forward and one toe facing backward. Ospreys and owls are the exception, with two toes facing each way. Each toe ends in a needle-sharp talon.

Most raptors dive toward their prey from the air and stun it with a forward thrust of their outstretched legs. They grasp the prey with their long toes and use the sharp back talon to pierce and hold it. Rough scales on the skin of the foot also help the bird to get a firm grip on its prey. The bird then carries its kill back to its nest or to a safe perch where it can eat without being disturbed.

LEFT: *A golden eagle has dark golden feathers on the nape of its neck.*

BELOW: *A golden eagle's foot is nearly the size of a human hand.*

RAPTORS USE their strong, sharp beaks to pull apart their prey and break it into pieces that are small enough to eat. Some predatory birds migrate long distances without stopping to eat, relying instead on stored body fat for energy. Most raptors, however, hunt for food often as they travel. They eat small mammals, birds, reptiles, fish, and, in some cases, insects. Their diets vary depending on their species and where they live.

Like most other birds, a raptor has a pouch inside its neck, called a crop, where it stores food. When people at the trapping stations examine a bird, they feel its crop. If the crop is bulging, they know that the bird has eaten recently.

Migration is a stressful time requiring a great deal of energy, and some birds do not make it to the end of their journey. Many raptors die of starvation, especially if they must cross over areas where food is hard to find. Some do not survive severe storms or other extreme weather conditions. Other birds are injured in accidents, become victims of predators, or are killed by hunters. Although shooting of raptors is illegal, some people do it because they mistakenly believe that the birds are pests.

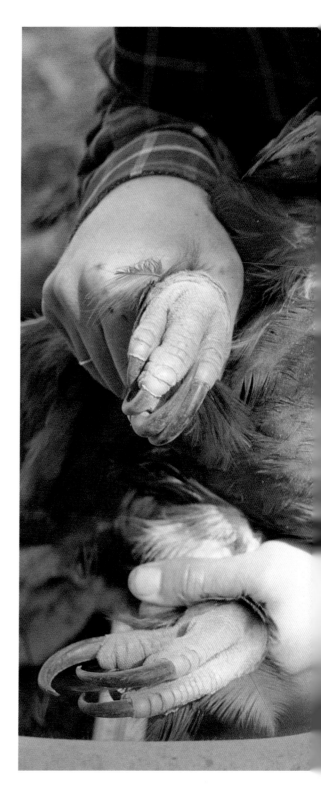

Measuring an eagle's beak is a two-person job.

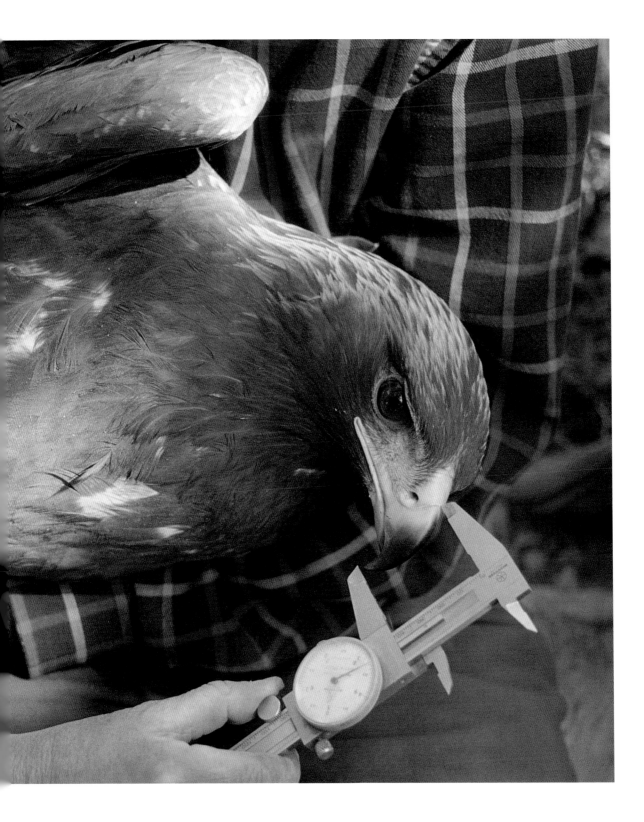

MORE THAN HALF OF THE BIRDS caught at the banding stations are juvenile birds that were hatched the previous spring. Unlike some species of birds, young raptors are fully grown by the time they leave the nest. They are the same size as adults, but the color of their feathers is usually different from that of adult birds. Often they are less distinctly marked—this may help them hide from predators while they are learning to hunt on their own. Birds that still have their juvenile feathers are called "immature," a reference only to their feathers, not to their size or development.

When raptors are about a year old they begin to molt, a process in which their old feathers are gradually replaced with new ones. The new feathers will have the coloring of an adult bird. Birds caught in the trapping stations that have a mixture of adult and juvenile feathers are called second-year birds. Eagles do not get their full adult plumage until they are several years old, but all other raptors have a complete set of adult feathers by the time they are a year and a half old.

Adult birds continue to molt once a year for the rest of their lives. Old feathers are easy to recognize because they are tattered at the edges and their color is faded. For birds older than one year, trappers keep a record of their old and new feathers.

The age of many hawks can also be estimated by looking at the color of their eyes. With buteos, the eyes darken as the bird grows older. The eyes of accipiters gradually change from yellow in first-year birds to ruby red in older adults. Falcons have dark eyes at all ages.

The worn brown tail feathers of this adult Cooper's hawk will soon be replaced by new gray feathers.

WHEN TRAPPERS HAVE COMPLETED all their measurements and observations of a captured bird, it is time to let it go. Whether the newly banded bird is a tiny kestrel or a golden eagle, its release and return to freedom is an exciting moment as it flaps its wings and rises in the air. As soon as the bird is on its way, the trappers contact the observers on the top of the mountain by shortwave radio and tell them the species, sex, and age of the bird that was just released. That way the bird is not counted again as it flies over the top of the ridge.

Chances are that the released bird will live the rest of its life with no further contact with humans. Yet, the hope of the trappers is that at least some of the banded birds will be found. Bands are recovered when birds are trapped again, usually at another location, and when birds are injured or killed. Recovered bands help us learn more about where birds go, how long it takes them to get there, and how they may have died in the wild. Fewer than 1 percent of all birds banded each year in the Goshutes are found again.

The National Biological Survey keeps a record of all birds banded in the United States. If you ever find a banded bird, you should report the band number, when and where you found the bird, and its condition to the Bird Banding Lab, Patuxent Wildlife Research Center, Laurel, Maryland 20708.

Under the supervision of an experienced raptor handler, a young visitor to the Goshutes briefly holds a sharp-shinned hawk and then releases it.

BY THE BEGINNING OF NOVEMBER, snow covers the ground in the Goshute Mountains, and most migrating raptors are well on their way south. It is time for the hardworking HawkWatch team to pack up their gear and hike down to the base of the mountain. Many of the biologists and volunteers will return the following summer to start another season of observation and banding.

The busiest time for raptor migration in the Goshutes is in the fall, and that is when HawkWatch International operates its trapping and banding station. Although some of the birds pass by again on their way north in the spring, most follow other routes. Migrating birds often fly one route as they go south and use another as they return north.

Raptor migration observation sites like the one in the Goshutes provide unique opportunities for close contact with predatory birds. Whether you spend a season banding or counting birds or visit a raptor migration lookout site for a day, the experience of seeing these magnificent predatory birds is unforgettable. The more we can learn about hawks, eagles, and falcons, the better we will be able to preserve the wild places where they live and ensure that these graceful hunters will have the chance to continue their annual migratory journeys on the highways of the sky.

ABOVE: *A prairie falcon, wearing a wire tracking device, heads into the clouds.*

LEFT: *A male American kestrel*

Day-Flying Raptors of North America

Family Accipitridae

Hawks

Buteos
- broad-winged hawk
- common black-hawk
- ferruginous hawk
- gray hawk
- Harris' hawk
- red-shouldered hawk
- red-tailed hawk
- rough-legged hawk
- short-tailed hawk
- Swainson's hawk
- white-tailed hawk
- zone-tailed hawk

Accipiters
- Cooper's hawk
- northern goshawk
- sharp-shinned hawk

Eagles

- bald eagle
- golden eagle

Kites

- American swallow-tailed kite
- black-shouldered kite
- hook-billed kite
- Mississippi kite
- snail kite

Harriers

- northern harrier

Ospreys

- osprey

Family Falconidae

Falcons

- American kestrel
- aplomado falcon
- gyrfalcon
- merlin
- peregrine falcon
- prairie falcon

Caracaras

- crested caracara

WATCHING RAPTORS
A Few North American Locations Where You Can See Migrating Birds of Prey

EASTERN UNITED STATES
1. Massachusetts: Mount Tom (fall)
2. New York: Hook Mountain (fall)
3. New York: Braddock Bay, south shore of Lake Ontario (spring)
4. Pennsylvania: Hawk Mountain (fall)
5. New Jersey: Cape May Point, entrance to Delaware Bay (fall)

MIDWESTERN UNITED STATES
6. Michigan: Whitefish Point, eastern end of Lake Superior (spring)
7. Minnesota: Hawk Ridge, western end of Lake Superior (fall)
8. Texas: Rio Grande Valley National Wildlife Refuge (fall and spring)
9. Texas: Corpus Christi (fall)

WESTERN UNITED STATES
10. New Mexico: Manzano Mountains (fall)
11. New Mexico: Sandia Mountains (spring)
12. Arizona: Lipan Point on the south rim of the Grand Canyon (fall)
13. Nevada: Goshute Mountains (fall)
14. California: Point Diablo at the Golden Gate Headland (fall)
15. Oregon: Bonney Butte (fall)
16. Montana: Bridger Mountains (fall)

CANADA
17. Alberta: Mount Lorette, Canadian Rockies (fall)
18. Ontario: Hawk Cliff, between Lake Erie and Lake Ontario (fall)

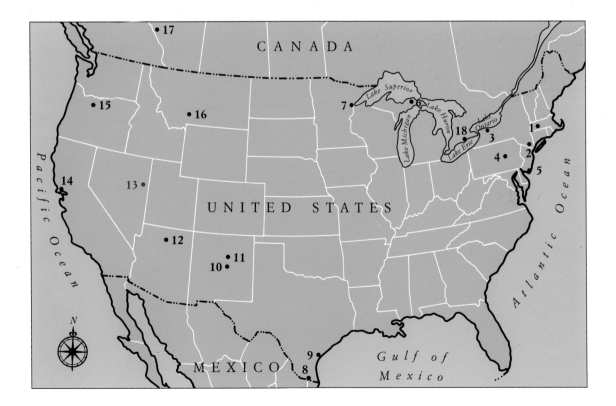

47

INDEX

accipiters, 15–17, 40, 46
Accipitridae, 15, 46

biologists, 7, 45
birds, predatory, *see* raptors
birds of prey, *see* raptors
blinds, 30–32, 34
buteos, 15, 40, 46

caracara, crested, 15, 21, 46
Cathartidae, 26

eagles, 7–8, 12, 15, 18, 26, 45, 46
   bald, 18, 46
   golden, 12, 18, 22–23, 36–39, 42, 46

Falconidae, 15, 21, 46
falcons, 5, 7–8, 12, 15, 17, 21, 24, 25, 26, 40, 45, 46
   aplomado, 21, 46
   peregrine, 10–11, 21, 24–25, 46
   prairie, 20–21, 45, 46
food chain, 8

goshawk, northern, 16–17, 28, 29, 46
Goshute Mountains, 5, 7, 12, 15, 17–18, 20–22, 24–26, 28–29, 31, 42, 45, 47
gyrfalcon, 21, 46

harrier, 15, 18, 19, 31, 46
hawks, 5–8, 11–12, 15, 17, 24–26, 28, 40, 45, 46
   broad-winged, 15, 24, 46
   Cooper's, 8–9, 12–13, 17, 40–41, 46
   ferruginous, 15, 46
   goshawk, 16, 29, 46
   marsh, *see* harrier
   red-tailed, 12, 14–15, 29, 46
   rough-legged, 15, 46
   sharp-shinned, 17, 18, 29, 34–35, 42–43, 46
   Swainson's, 15, 24, 46
   true, *see* accipiters
HawkWatch International, 5, 7, 28, 45
hawk-watching sites, 5, 12, 25–26, 28, 45, 47

kestrel, American, 2, 21, 28–29, 32, 42, 44–45, 46
kites, 15, 18, 46

merlin, 20–21, 46
migration, *see* raptor: migration; raptors: migratory

National Biological Survey, 42
North America, 5, 7, 12, 15, 18, 21, 22, 25, 28, 47

osprey, 15, 37, 46
owls, 8, 37

pesticides, 11
prey, 7–8, 11–12, 17, 21, 26, 35, 37–38

raptor,
   beaks, 7–8, 31, 38–39
   eyes, 8, 17, 28, 34–35, 40
   feathers, 36–37, 40–41
   feet, 8, 37
   migration, 5, 12, 17, 22, 24–26, 28, 38, 45
   tails, 14–15, 17–18, 21, 40–41
   talons, 7–8, 31, 33, 37
   weight, 7, 34–35
   wings, 7, 15, 17–18, 21, 24, 42–43
raptors, 5, 7–8, 11–12, 18, 22, 24–26, 28–29, 31, 34–35, 37–38, 40, 45, 46–47
   banding of 5, 7, 12, 17, 28, 32–33, 40, 42, 45
   counting of, 7, 11, 26, 28, 45
   identification of, 26
   measuring of, 7, 32–34, 42
   migratory, 5, 8, 11–12, 22, 24–26, 38
   observation of, 5, 7, 12, 26–27, 42, 45
   releasing of, 7, 32, 42–43
   trapping of, 5, 7, 12, 17, 28–32, 38, 40, 42, 45

scientists, 7–8, 11, 15, 22, 28, 35

vultures, turkey, 26